TIPS AND TRICKS TO BECOME BETTER AT MATH

Many individuals accept one is brought into the world with the ability to be great at math while others are simply not. In any case, many examinations have demonstrated there's no inborn numerical capacity; everybody can become capable in math in the event that they set forth the energy and time. While it very well may be overwhelming, math is a sort of language that'd constantly have a response. All things considered, numbers don't lie. Hoping to gain proficiency with the Tips And Tricks To Become Better At Math. Peruse this book to figure out the best methodologies to further develop your number related abilities.

Chapter by chapter guide

Comprehend The Topic Before Moving On To Another

Be Math-Proficient!

1. Separate Complex Problems

2. Ace The Basic Math Skills

3. Comprehend The Topic Before Moving On To Another

4. Know The Importance Of Number Sense

5. Have A Regular And Consistent Practice

6. Lay out A Routine

7. Center Around Understanding New Concepts

8. Make A Practice Math Test

9. Record Each Math Step

10. Practice Mental Math

11. Download Useful Math Apps

12. Apply Math Problems In Real-World Challenges

13. Consider Enrolling In Private Math Tutorials

End

Comprehend The Topic Before Moving On To Another

Math is like perusing since, in such a case that you don't have the foggiest idea how your letter sounds, you get no opportunity of expressing words, so you'd not be able to peruse an expression or a sentence.

All number related courses follow a recommended request since every point expands on the former one. In the event that you're encountering trouble with a specific subject, work on it until you grasp it and settle it. Try not to skirt different subjects since it could hamper your advancement as you go on.

Watch video instructional exercises, go to math conversations, practice test activities, or even buy a bunch of books introducing alternate methods of critical thinking. In any case, don't continue on toward the following point on the off chance that you actually find the subject muddled. Assuming you do, you'll get significantly more bombshell and you'll unquestionably start to lose trust.

Be Math-Proficient!

Knowing how to apply fundamental math is basic for ordinary living, future profession, and business potential open doors. Critical investigations on one's capacity to foster numerical abilities show it's equivalent to some other abilities one is equipped for acquiring, so the expectation isn't lost in the event that you see as the subject testing.

Here are a few quick tips and deceives that can assist you with succeeding in math:

1. Separate Complex Problems

There's consistently a simpler method for tackling an issue. Get to the core of any perplexing numerical statement by distinguishing reasonable systems, disconnecting them, and work on settling the unexplored world. Attempt to simplify terms by separating them to the most essential structure.

For instance, to take the proportion of a point, search for the least demanding technique to do this. Estimating the worth of a point inside a triangle is straightforward while utilizing the legitimate procedures. Whenever you've excelled at viewing as the obscure, you'll know how to find a unidentified variable utilizing right points, advantageous points, and correlative points.

2. Ace The Basic Math Skills

Estimations including numbers, sizes, or different measures are viewed as fundamental numerical abilities. These abilities incorporate the basics like expansion and deduction as well as further developed math thoughts based on them. Having great numerical capacities will help you in school advancing as well as in daily existence.

Fundamental numerical abilities are the accompanying:

Expansion, deduction, augmentation, division

To settle a numerical issue, understanding these four operations is significant. You'll have to know how to add, take away, duplicate, and gap straightforward numbers to you to perform anything math-related.

Parts and decimals

Decimals are a part of an entire number while divisions are the mathematical portrayal of the decimal. At the point when parts are presented, it's critical to begin with lower numbers like 1/4 as well as to follow the proportions (1:4, 1:25). An effective method for further developing your math capacities is to figure out how to add, deduct, duplicate, and gap parts.

Rate

It's a part of the entire or a specific sum for each 100. This is utilized while you're chipping away at tasks like sorting out the markdown, nourishment esteem, registering deals charge, loan cost in your bank reserve funds, and that's only the tip of the iceberg.

Visual portrayal of information

Numbers are ordinarily introduced in different ways, so they can be better perceived. To have the option to peruse and decipher patterns, you should have essential number-crunching abilities. Having the capacity to get a superior handle of hidden information, the capacity to decipher the pattern line, data of interest, and tomahawks is significant. Furthermore, it'll help you in making your diagrams and outlines, making it more straightforward to impart your thoughts.

Visual portrayal of information

Tackling the unexplored world

Knowing the worth of an obscure variable is a typical issue in polynomial math. At the end of the day, in the event that Juan needs to procure USD$600 this month and he gets USD$20 each time he strolls a canine for one of his neighbors, he'll have USD$800 toward the month's end. He'd be persuaded to sort out the number of strolls that it'd take to bring in how much cash he wanted. Without even a trace of more convoluted science, Juan could think of a basic strategy to get the obscure variable in this situation (20x = 600) and afterward settle for the unexplored world. Likewise, an obscure worth in an extent can be determined by cross-result of two divisions. You need to isolate the variable from different amounts to get its worth. An extent number cruncher could be a convenient device to track down the obscure worth in an extent.

3. Comprehend The Topic Before Moving On To Another

Math is like perusing since, in such a case that you don't have the foggiest idea how your letter sounds, you get no opportunity of expressing words, so you'd not be able to peruse an expression or a sentence. All numerical courses follow an endorsed request since every point expands on the previous one. In the event that you're encountering trouble with a specific subject, work on it until you fathom it and tackle it. Try not to skirt different themes since it could hamper your advancement as you go on. Watch video instructional exercises, go to math conversations, practice test activities, or even buy a bunch of books introducing alternate methods of critical thinking. In any case, don't continue on toward the following point on the off chance that you actually find the subject muddled. On the off chance that you do, you'll get much more surprise and you'll absolutely start to lose trust.

4. Know The Importance Of Number Sense

Many individuals retain math to gain proficiency with the idea. This might likewise show they'll retain the augmentation table. Be that as it may, when you experience an unexpected flood of nervousness while taking a test, you can fail to remember anything you've recently recollected.

All things considered, having a strong comprehension of number sense is more worthwhile. For instance, you might utilize 10×8 to appreciate what 8×9 is. Taking away eight from 80 will lead you to 72 on the grounds that you're looking for gatherings of eight as opposed to ten.

In the event that you've fostered a decent number sense, you can utilize that expertise to make calculations more straightforward and gain trust in moving toward different situations of a similar sort. You can distinguish on the off chance that a response is satisfactory with great number sense and you can lay out replies prior to utilizing calculation.

5. Have A Regular And Consistent Practice

One more tip to turn out to be great at math is to rehearse your numerical abilities as frequently as could be expected. This implies you need to invest some energy taking care of numerical statements every day. The more you practice, the better you'll turn into.

On the off chance that you're experiencing difficulty with math at school, you can get an additional assistance by conversing with an instructor, discovering some free internet based math coaching locales, or taking some useful number related instructional classes. Math guides can offer you tips to reprieve down complex conditions and tell you the best way to take care of issues.

6. Lay out A Routine

To turn out to be better at math is to have a daily practice in taking care of issues and dominating the idea. In the event that you figure out how to tackle, it becomes more straightforward for you to settle a similar condition sometime later. This will save you time and work over the long haul. This is particularly obvious in the event that you have an enormous issue to tackle.

On the off chance that you can get into a decent daily schedule with your number related learning, you'll not need to stress over rehashing a similar issue again and again in light of the fact that you as of now ace it. You'd see themes and patterns as you rehash specific issues.

7. Center Around Understanding New Concepts

To effectively finish numerical issues, you can recollect conditions and rules, yet this doesn't infer you've perceived the standards behind the thing you're doing. Subsequently, it turns out to be more hard to address issues and to learn anything new. The additional time you take to zero in on seeing new ideas, the more capable you'll be in your number-crunching abilities.

Beginning to manage other numerical statements straight away could bring about disappointment and disarray. Continuously focus in classes or math instructional exercises and search for the most straightforward numerical subtleties and recipes that will help you in your examinations.

Furthermore, it's additionally strongly suggested you invest energy rehearsing numerical questions your teachers offer prior to completing any tasks. Realizing this will offer you a chance to completely understand what you've realized.

8. Make A Practice Math Test

Math concentrate on meetings ought to consolidate taking care of issues and evaluating one's comprehension by going through models and practice questions.

The best procedure to concentrate on math is to utilize practice tests. Reproduce a test you could experience from here on out or gather past training questions and build a counterfeit test for you to finish. This will give you an amazing chance to settle genuine test inquiries ahead of time.

9. Record Each Math Step

A great many people will attempt to take care of numerical questions in their cerebrums. Nonetheless, convoluted numerical questions would require you to go through a few sorts of data and make numerous derivations prior to showing up at an answer. Endeavoring to do all that immediately can make you be confounded and baffled.

Work through a numerical statement bit by bit while recording it to guide you through the settling system. As you become further developed, this approach will become helpful as you work through progressively troublesome number related issues, which empowers you to go through bit by bit. In the event that you commit an error, you can return and take a gander at your moves toward gain from them.

10. Practice Mental Math

Mental computation execution gives significant benefits in certain circumstances. Mental number juggling works on one's intellectual ability. The number to measure the climate around you works on legitimate and natural reasoning. Fortifying your establishment by rehearsing mental estimations will assist you with acquiring a more profound comprehension of additional troublesome ideas.

For instance, working out the amount to tip a server at an eatery is a straightforward number-crunching issue numerous people can't tackle without the utilization of a mini-computer. Via preparing your cerebrum to answer basic numerical statements, you can save time in conditions, for example, these.

Observe, mental math is unique in relation to retaining numerical data, for example, augmentation tables. It's simplified with a groundwork of recalled answers for rudimentary numerical statements, however executing it in your cerebrum takes both learned information and control of numbers and tasks to tackle. This mix of capacity and memory empowers you to take care of undeniably more muddled issues than can be addressed with promptly retained information.

11. Download Useful Math Apps

There's an obvious association among math and innovation. One method for ensuring you in all actuality do well in math is to utilize the right learning apparatuses and programs. Since many are now becoming stuck to their mobile phones, it just checks out they'd need to utilize a math application that instructs them.

The iPhone and Android working frameworks have numerous famous numerical projects for understudies to download. Exploit free math applications as you begin to feel more certain about your abilities while rehearsing the essential ideas during your available energy. Likewise, these stages are profitable as the majority of them give replies on the best way to tackle the issues.

12. Apply Math Problems In Real-World Challenges

To make math more significant, apply them to true difficulties. Certain individuals view math as both dynamic and inconsequential to the world. Regardless of whether this is valid dependent upon some degree, it doesn't need to be seen like that. For example, the Pythagorean Theorem alludes to the connections between various measured structures, so attempt to apply it in regular stuff including triangles. Attempt to further develop your number related abilities by sorting out ways of applying anything that you figure out how to your own life.

13. Consider Enrolling In Private Math Tutorials

Math coaching can likewise cause you to figure out various points in a lot more prominent profundity. For instance, a few understudies who come up short may not understand the fundamental issue lies in the feeble embrace of essential ideas. Through the help of a guide, one will actually want to perceive how different subjects are connected with one another. This will assist them with working on how they might interpret math, making their abilities more effective.

Math can already be hard for what it's worth, however with the direction of an accomplished number related mentor, you'll have the chance to learn at your speed. So on the off chance that you have battles in math, confidential math coaching can give the assets you should find success. Whether you're searching for a method for making you more serious or you basically need to approach the devices you want to succeed, then confidential math coaching can give what you want.

End

If you have any desire to do well in math, then you ought to think consistently and not simply cursorily. Whenever you've been presented to this perspective, you'll in all probability treat the subject more in a serious way. Despite the fact that there are numerous ways you can improve at math, there could be no silver projectile. You need to manage slowly but surely, however the more you practice, the better you'll be grinding away.

Utilize the ideas above to turn out to be better at math, and you'll have more certainty to apply ideas, in actuality, circumstances.

www.ingramcontent.com/pod-product-compliance
Lightning Source LLC
Chambersburg PA
CBHW080438220526
45465CB00009B/3341